The
Usefulness
of Useless
Knowledge

Princeton University Press
Princeton and Oxford

The Usefulness of Useless Knowledge

ABRAHAM FLEXNER

With a companion essay by
ROBBERT DIJKGRAAF

Published by Princeton University Press,
41 William Street, Princeton, New Jersey 08540

In the United Kingdom: Princeton University Press,
6 Oxford Street, Woodstock, Oxfordshire OX20 1TR

press.princeton.edu

Jacket design by Amanda Weiss

Library of Congress Control Number 2016960697
ISBN 978-0-691-17476-1

British Library Cataloging-in-Publication Data is available

This book has been composed in Miller

Printed on acid-free paper. ∞

Printed in the United States of America

1 3 5 7 9 10 8 6 4 2

Contents

The
Usefulness
of Useless
Knowledge

The World of Tomorrow

ROBBERT DIJKGRAAF

On April 30, 1939, under the gathering storm clouds of war, the New York World's Fair opened in Flushing Meadows, in Queens. Its theme was *The World of Tomorrow*. Over the next eighteen months, nearly forty-five million visitors would be given a peek into a future shaped by newly emerging technologies. Some of the displayed innovations were truly visionary. The fair featured the first automatic dishwasher, air conditioner, and fax machine. The live broadcast of President Franklin Roosevelt's opening speech introduced America to television. Newsreels showed Elektro the Moto-Man, a seven-foot tall, awkwardly moving aluminum robot that could speak by playing 78-rpm records, smoke a cigarette, and play with his robot dog Sparko. Other attractions, such as a pageant featuring magnificent steam-powered locomotives, could be better characterized as the last gasps of the world of yesterday.

Albert Einstein, honorary chair of the fair's science advisory committee, presided over the official illumination ceremony, also broadcast live on television. He spoke to a huge crowd on the topic of cosmic rays, highly energetic subatomic particles bombarding the Earth from outer space. The event has been described as a comedy of errors. Einstein's talk could hardly be understood as the amplification system soon broke down. And the opening act—the capture of ten cosmic rays—ended with a spectacular debacle. The particles were transported by telephone line from the Hayden Planetarium in Manhattan to the fairgrounds in Queens, where bells and lights signaled their arrival. But when the tenth ray was captured, a power failure occurred to the great disappointment of the audience, which soon decamped. As the *New York Times* reported the next day, "The crowd dropped science in favor of a spectacle that they could applaud."

Two scientific discoveries that would soon dominate the world were absent at the World's Fair: nuclear energy and electronic computers.

Remarkably, the very beginnings of both technologies could be found at an institution that had been Einstein's academic home since 1933: the Institute for Advanced Study in Princeton, New Jersey. The Institute was the brainchild of its first director, Abraham Flexner. Intended to be a "paradise for scholars" with no students or administrative duties, it allowed its academic stars to fully concentrate on deep thoughts, as far removed as possible from everyday matters and practical applications. It was the embodiment of Flexner's vision of the "unobstructed pursuit of useless knowledge," which would only show its use over many decades, if at all.

However, the unforeseen usefulness came much faster than expected. By setting up his academic paradise, Flexner unintentionally enabled the nuclear and digital revolutions. Among his first appointments was Einstein, who would follow his speech at the World's Fair with his famous letter to President Roosevelt in

August 1939, urging him to start the atomic bomb project. The breakthrough paper by Niels Bohr and John Wheeler on the mechanism of nuclear fission appeared in the *Physical Review* on September 1, 1939, the same day World War II started.

Another early Flexner appointee was the Hungarian mathematician John von Neumann, perhaps an even greater genius than Einstein, of almost extraterrestrial brilliance. Von Neumann was one of the "Martians," an influential group of Hungarian scientists and mathematicians that also included Edward Teller, Eugene Wigner, and Leo Szilard, the physicist who helped draft Einstein's letter to Roosevelt. A well-told story in physics is that when a frustrated Enrico Fermi asked where were the highly exceptional and talented aliens that were meant to find Earth, an impish Szilard replied, "They are among us, but they call themselves Hungarians."

Von Neumann's early reputation was based on his work in pure mathematics and the foundations of quantum theory. Together with the American logician Alonzo Church, he made Princeton a center for mathematical logic in the 1930s, attracting such luminaries as Kurt Gödel and Alan Turing. Von Neumann was fascinated by Turing's abstract idea of a universal calculating machine that could mechanically prove mathematical theorems. When the nuclear bomb program required large-scale numeric modeling, von Neumann gathered a group of engineers at the Institute to begin designing, building, and programming an electronic digital computer—the physical realization of Turing's universal machine. As von Neumann observed in 1946, "I am thinking about something much more important than bombs. I am thinking about computers."

In his spare time, von Neumann directed his team to focus these new computational powers on many other problems aside from weapons. With meteorologist Jule Charney, he made the first numerical weather prediction in 1949—technically it was a "postdiction," since at that time it took forty-eight hours to predict tomorrow's weather. Anticipating our present climate-change reality, von Neumann would write about the study of the Earth's weather and climate: "All this will merge each nation's affairs with those of every other, more thoroughly than the threat of a nuclear or any other war may already have done."

A logical machine that can prove mathematical theorems or a highly technical paper on the structure of the atomic nucleus may seem to be useless endeavors. In fact, they played important roles in developing technologies that have revolutionized our way of life beyond recognition. These curiosity-driven inquiries into the foundations of matter and calculation led to the development of nuclear arms and digital computers, which in turn permanently upset the world order, both militarily and economically. Rather than attempting to demarcate the nebulous and artificial distinction between "useful" and "useless" knowledge, we may follow the example of the British chemist and Nobel laureate George Porter, who spoke instead of applied and "not-yet-applied" research.

Supporting applied and not-yet-applied research is not just smart, but a social imperative. In order to enable and encourage the full cycle of scientific innovation, which feeds into society in numerous important ways, it is more productive to think of developing a solid portfolio of research in much the same way as we approach well-managed financial resources. Such a balanced portfolio would contain predictable and stable short-term investments, as well as long-term bets that are intrinsically more risky but can potentially earn off-the-scale rewards. A healthy and balanced ecosystem would support the full spectrum of scholarship, nourishing a complex web of interdependencies and feedback loops.

However, our current research climate, governed by imperfect "metrics" and policies, obstructs this prudent approach. Driven by an ever-deepening lack of funding, against a background of economic uncertainty, global political turmoil, and ever-shortening time cycles, research criteria are becoming dangerously skewed toward conservative short-term goals that may address more immediate problems but miss out on the huge advances that human imagination can bring in the long term. Just as in Flexner's time, the progress of our modern age, and of the world of tomorrow, depends not only on technical expertise, but also on unobstructed curiosity and the benefits—and pleasures—of traveling far upstream, against the current of practical considerations.

* * *

Who was Abraham Flexner, and how did he arrive at his firm beliefs in the power of unfettered scholarship? Born in 1866 in Louisville, Kentucky, Flexner was one of nine children of Jewish immigrants from Bohemia. In spite of sudden economic hardship—the Flexners lost their business in the panic of 1873—and with the help of his older brother Jacob, Abraham was able to attend Johns Hopkins University, arguably the first modern research university in the United States. Flexner's exposure to the advanced opportunities at Johns Hopkins, which were comparable to those at leading foreign universities, permanently colored his views. He remained a lifelong critic and reformer of teaching and research. After obtaining his bachelor's degree in classics in just two years, he returned to Louisville, where he started a college preparatory school to implement his revolutionary ideas based on a deep confidence in the creative powers of the individual and an equally deep distrust of the ability of institutions to foster such talent.

Flexner first rose to public attention in 1908 with his book *The American College: A Criticism* with a strong appeal for hands-on teaching in small classes. His main claim to fame was his 1910 "bombshell report," commissioned by the Carnegie Foundation, on the state of 155 medical schools in North America, branding many of them as frauds and irresponsible profit machines that withheld from students any practical training. He didn't hesitate to label institutions as disgraceful, shameful, or even fictional. Chicago was characterized as "the plague spot of the country." The effectiveness of the *Flexner Report* is the stuff of advisory committee dreams. It led to the closure of almost half of the medical schools and the wide reform of others, starting the age of modern biomedical teaching and research in the United States.

Flexner's efforts and vision led to his joining the General Education Board of the Rockefeller Foundation in 1912, lending him added stature and resources as an influential force in higher education and philanthropy. He soon became its executive secretary, a position he held until his retirement in 1927. It was in this capacity that the ideas underlying his essay "The Usefulness of Useless Knowledge" would form. It would finally be published in *Harper's* magazine in October 1939, but it began as a 1921 internal memo prepared for the board. In the 1920s, Flexner carefully studied institutions of higher education across Europe, from the ancient colleges of England and France to the modern research universities and institutes of Germany, with their strong links to industry. An opportunity to give the 1928 Rhodes Trust Memorial lectures in Oxford while in residence at All Souls College crystalized his ideas about the future of universities and research institutions. An expansion of his well-received three lectures was published as *Universities: American, English, German* (Oxford 1930). The Great Depression and the political unrest leading to another world war in the thirties would only sharpen his arguments for the need for independent scholarship.

Flexner was given the opportunity to put his lofty vision into practice when he was approached in 1929 by representatives of Louis Bamberger and his sister Caroline Bamberger Fuld. The Bambergers had sold their massive, eponymous Newark department store to Macy's a few weeks before the Wall Street crash, leaving them with a large fortune. Their original intent was to found a medical institution without racial, religious, or ethnic biases, but Flexner convinced the benefactors to set up an institute exclusively dedicated to unrestricted scholarship. In 1930, he became the founding director of the Institute for Advanced Study in Princeton.

The mission and vision of the Institute expanded drastically with the turn of events in Europe. The first scholars, including Einstein, arrived in Princeton in 1933, just when Hitler came to power and his draconian laws prompted an exodus of Jewish scientists from Germany. Flexner worked with his brothers Simon and Bernard and the Rockefeller Foundation to bring as many scholars as possible to the United States. This influx of European talent would dramatically alter the global balance of knowledge. In May 1939, Flexner wrote in his last annual director's report, "We are living in an epoch-making time. The center of human culture is being shifted under our very eyes. . . . It is now being unmistakably shifted to the United States. . . . Fifty years from now the historian looking backward will, if we act with courage and imagination, report that during our time the center of gravity in scholarship moved across the Atlantic Ocean to the United States." Flexner did as much as anyone to make this happen. When Abraham Flexner died in 1959 at age 92, his obituary appeared on the front page of the *New York Times* along with an editorial concluding, "No other American of his time has contributed more to the welfare of this country and of humanity in general."

* * *

It was Flexner's lifelong conviction that human curiosity, with the help of serendipity, was the only force strong enough to break through the mental walls that block truly transformative ideas and technologies. He believed that only with the benefit of hindsight could the long arcs of knowledge be discerned, often starting with unfettered inquiry and ending in practical applications.

Flexner articulates well the effect of the groundbreaking investigations into the nature of electromagnetism by Michael Faraday and James Clerk Maxwell—recall that the year 1939 saw the introduction of FM radio and television to the United States. Remarkably, on the wall of Einstein's home office hung small portraits of these two British physicists. There is a famous, but most likely apocryphal, anecdote that when William Gladstone, then the Chancellor of the Exchequer, visited the laboratory of Faraday in the 1850s and inquired what practical good his experiments in electricity would bring the nation, Faraday answered, "One day, Sir, you may tax it." The equations were never patented, but it is hard to think of any human endeavor today that doesn't make use of electricity or wireless communication. Over a century and a half, almost all aspects of our lives have literally been electrified.

In the same way, in the early twentieth century the study of the atom and the development of quantum mechanics were seen as a theoretical playground for a handful of often remarkably young physicists—one spoke of *Knabenphysik*, boys' physics—with little immediate consequences. The birth of quantum theory was long and painful. The German physicist Max Planck described his revolutionary thesis, first proposed in 1900, that energy could only occur in packets or "quanta" as "an act of desperation." In his words, "I was willing to make any offer to the principles in physics that I then held." His gambit played out very well. Without quantum theory, we wouldn't understand the nature of any material, including its color, texture, and chemical and nuclear properties. These days, in a world totally dependent on microprocessors, lasers, and nanotechnology, it has been estimated that 30 percent of the U.S. gross national product is based on inventions made possible by quantum mechanics. With the booming high-tech industry and the expected advent of quantum computers, this percentage will only grow. Within a hundred years, an esoteric theory of young physicists became a mainstay of the modern economy.

It took nearly as long for Einstein's own theory of relativity, first published in 1905, to be used in everyday life in an entirely unexpected way. The accuracy of the global positioning system (GPS), the space-based navigation system that provides location and time information in today's mobile society, depends on reading time signals of orbiting satellites. The presence of the Earth's gravitational field and the movement of these satellites cause clocks to speed up and slow down, shifting them by thirty-eight milliseconds a day. In one day, without Einstein's theory, our GPS tracking devices would be inaccurate by about seven miles. Again, a century of free-flowing thinking and experimentation led to a technology that literally guides us every day.

The path from exploratory blue-sky research to practical applications is not one-directional and linear, but rather complex and cyclic, with resultant technologies enabling even more fundamental discoveries. Take, for example, superconductivity, the phenomenon discovered by the Dutch physicist Heike Kamerlingh Onnes in 1911. Certain materials, when cooled down to ultralow temperatures, turn out to conduct electricity without any resistance, allowing large electric currents to flow at no costs. The powerful magnets that can be so constructed have led to many innovative applications, from the maglev transport technology that allows trains to travel at very high speeds as they levitate through magnetic fields to the fMRI technology used to make detailed brain scans for diagnostic and treatment purposes.

Through these breakthrough technologies, superconductivity has in turn pushed the frontiers of basic research in many directions. High-precision scanning has made possible the flourishing field of present-day neuroscience, which is probing the deepest questions about human cognition and consciousness. Superconductivity is playing a crucial role in the development of quantum computers and the next revolution in our computational powers with unimaginable consequences. And in fundamental physics, it has produced the largest and strongest magnets on the planet, buried a hundred meters underground in the twenty-seven-kilometer-long ring of the Large Hadron Collider, the particle accelerator built in the CERN laboratory in Geneva. The resulting 2012 discovery of the Higgs boson was the capstone that completed the Standard Model of particle physics, enabling physicists to further probe and unravel the mysteries of the universe. Remarkably, the deep understanding of the Higgs particle is itself based on the theory of superconductivity. There is therefore an evident route from the discovery of superconductivity to the discovery of the Higgs particle one century later. But it is hardly a straight one, going through many loops.

The life sciences provide perhaps the richest source of powerful practical implications of fundamental discoveries. One of the least known success stories in human history is how over the past two and a half centuries advances in medicine and hygiene have tripled life expectancy in the West. The discovery of the double helical structure of DNA in 1953 jump-started the age of molecular biology, unraveling the genetic code and the complexity of life. The advent of recombinant DNA technology in the 1970s and the completion of the Human Genome Project in 2003 revolutionized pharmaceutical research and created the modern biotech industry. Currently, the CRISPR-Cas9 technology for gene editing allows scientists to rewrite the genetic code with unbounded potential for preventing and treating diseases and improving agriculture and food security. We should never forget that these groundbreaking discoveries, with their immense consequences for health and diseases, were products of addressing deep basic questions about living systems, without any thoughts of immediate applications.

* * *

Flexner's perspective on the "usefulness of useless knowledge" has only gained in substance and breadth since his time. First and foremost, as Flexner argues so elegantly, basic research clearly advances knowledge in and of itself. Fundamental inquiry moves exploration as far up to the headwaters as possible, producing ideas that slowly and steadily turn into concrete applications and further studies. As it is often stated, knowledge is the only resource that increases when used.

Second, pathbreaking research leads to new tools and techniques, often in unpredictable and indirect ways. A remarkable, late-twentieth-century example of such a fortuitous outgrowth was the development of automatic information-sharing software, introduced as the World Wide Web in 1989. What began as a collaboration tool for thousands of particle physicists working at the CERN particle accelerator laboratory entered the public domain in 1993, unleashing the power of the Internet to the masses and facilitating large-scale communication around the globe. To store and process the vast amount of data produced in the same particle experiments, so-called grid and cloud computing were developed, linking computers in huge virtual networks around the world. These cloud technologies now drive many Internet business applications, from services and shopping to entertainment and social media.

A third attribute is the attraction of curiosity-driven research to the very best minds in the world. Young scientists and scholars, drawn to the intellectual challenges of fundamental questions, are trained in completely new ways of thinking and using technology. Once these skills carry over to society, they can have transformative effects. For example, scientists who have learned to capture complex natural phenomena in elegant mathematical equations apply these techniques to other branches of society and industry, such as in the quantitative analysis of financial and social data.

Fourth, much of the knowledge developed by basic research is made publicly accessible and so benefits society as a whole, spreading widely beyond the narrow circle of individuals who, over years and decades, introduce and develop the ideas. Fundamental advances in knowledge cannot be owned or restricted by people, institutions, or nations, certainly not in the current age of the Internet. They are truly public goods.

Finally, one of the most tangible effects of pathbreaking research appears in the form of start-up companies. The new industrial players of the past decades show how powerful technologies are in generating commercial activities. It is estimated that more than half of all economic growth comes from innovation. Leading information technology and biotech industries can trace their success directly to the fruits of fundamental research grown in the fertile environments around research universities such as in Silicon Valley and the Boston area, often infused by generous public investments. MIT estimates that it has given rise to more than thirty thousand companies with roughly 4.6 million employees, including giants such as Texas Instruments, McDonnell Douglas, and Genentech.

The two founders of Google worked as graduate students at Stanford University on a project supported by the Digital Libraries Initiative of the National Science Foundation—possibly the government grant with the highest payoff ever.

Flexner was not the first to argue for the power of curiosity and imagination. In the "Usefulness of Useless Knowledge," he writes, "Curiosity, which may or may not eventuate in something useful, is probably the outstanding characteristic of modern thinking. It is not new. It goes back to Galileo, Bacon, and to Sir Isaac Newton, and it must be absolutely unhampered."

The role of imagination in science had an early advocate in the Dutch chemist Jacobus Henricus van 't Hoff, the first Nobel laureate in chemistry. In 1874, then only twenty-two years old, he issued a brief pamphlet introducing the idea that molecules could only be properly understood by considering their three-dimensional spatial form—literally opening the eyes of chemistry. Not everyone immediately embraced his radical insight. One of the great figures of the time, the German chemist Hermann Kolbe, editor of the prominent *Journal für praktische Chemie*, was merciless in his criticism: "A certain Dr. J.H. van 't Hoff, who works at the Veterinary School in Utrecht, apparently does not believe in exact chemical research. He finds it more comfortable to mount Pegasus (probably borrowed from the Veterinary School) and, after a bold flight to the chemical Parnassus, simply pronounce how the positions of the atoms in space appeared before him."

These words of criticism hit hard, even though they aroused a lot of international interest—even then, there was no such thing as bad publicity. Four years later, when Van 't Hoff was appointed professor at the University of Amsterdam, he fiercely defended the role of creativity in his inaugural speech *Imagination in Science* (Bazendijk, Rotterdam, Netherlands 1878).

As a good chemist, he approached the subject experimentally, collecting the biographies of two hundred famous scientists and looking carefully for any interest in the arts and literature. He quotes approvingly from a letter of Faraday: "Do not suppose that I was a very deep thinker, or was marked as a precocious person. I was a very lively imaginative person, and could believe in the *Arabian Nights* as easily as in the *Encyclopaedia*."

Van 't Hoff reports indications of powerful imagination in fifty-two of his two hundred subjects, supporting his main thesis to a very large extent. Incidentally, eleven of these cases, including such intellectual giants as Newton, Leibniz, and Descartes, show symptoms of "pathological" imagination with a tendency for superstition, hallucination, spiritism, alchemy, and metaphysical speculation.

Into the twentieth century and beyond, imagination has been a driving force of successful scientists and scholars across disciplines. Einstein famously said, "Imagination is more important than knowledge. For knowledge is limited to all we now know and understand, while imagination embraces the entire world, and all there ever will be to know and understand." The particle physicist Richard Feynman, who turned into a modern cultural icon for ingenuity, even featured in advertisements for Apple computers, nicely captured how science combines imagination with rigorous methods: "Scientific creativity is imagination in a straitjacket."

* * *

While the big-picture arguments for blue-sky research driven by curiosity and imagination are as timely and relevant as ever, much has happened since the publication of Flexner's essay. The seminal contributions of scientists during the war years, as exemplified by the Manhattan

Project, led to the broad realization that basic research is crucial to the survival of the nation and the world. As director of the Office of Scientific Research and Development during World War II, Vannevar Bush produced a report in 1945, at President Roosevelt's request, that captured and communicated that insight. Bush's *Science, the Endless Frontier* ushered in a postwar boom in public funding of basic science, first in the United States and soon across the Western world. Remarkably, despite the obvious immediate need for weapons research, the intrinsic cultural value of science and scholarship was consistently emphasized. As physicist Robert Wilson would testify in a 1969 congressional hearing about the possible Cold War use of the Fermilab particle accelerator, "This new knowledge has all to do with honor and country, but it has nothing to do directly with defending our country, except to help make it worth defending." During this same period, the American liberal arts tradition in education was revitalized, embracing the humanities as an anchor for the fundamental values for which World War II had been fought.

As a consequence, the postwar decades saw an unprecedented worldwide growth of science, including the creation of funding councils like the National Science Foundation and massive investments in research infrastructure. Another crucial impetus came with the launch of a basketball-sized Soviet space vehicle on October 4, 1957. Sputnik was a watershed moment for American education and research. It reformed the science curriculum with an emphasis on hands-on experiments, led to the creation of NASA and the space race, set up the advanced research agency DARPA within the Department of Defense, and substantially increased research funding for science and engineering. The present age of microelectronics and the Internet can trace its beginnings directly to the Sputnik effect.

Recent decades have seen a marked retrenchment from that positive trend. One can argue that the state of scholarship has now reached a critical stage that in many ways mirrors the crisis that Flexner addressed. Steadily declining public funding is currently insufficient to keep up with the expanding role of the scientific enterprise in a modern knowledge-based society. The U.S. federal research and development budget, measured as a fraction of the gross domestic product, has steadily declined, from a high of 2.1 percent in 1964, at the height of the Cold War and the space race, to currently less than 0.8 percent. (Note that roughly half of that budget has remained defense-oriented.) The budget for the National Institutes of Health, the largest funder of medical research in the United States, has fallen by 25 percent over the past decade.

On top of this, industry, driven by short-term shareholder pressure, has been steadily decreasing its research activities, transferring that responsibility largely to public funding and private philanthropy. A committee of the U.S. Congress found that in 2012 business only provided 6 percent of basic research funding, with the lion's share—53 percent—shouldered by the federal government and the remainder coming from universities and foundations. It is difficult to imagine a corporation today following the example of the famous Bell Laboratories, whose scientists were awarded eight Nobel Prizes for basic research that led to the development of the transistor and the laser and the discovery of cosmic microwave background radiation, a signature of the Big Bang theory of the beginning of the universe, among other developments. These days, the added burden of industrial research is crowding out basic research at many universities. Meanwhile, governments are increasingly directing research funding to tackle important societal challenges, such as the transition to clean sustainable energy, battling climate change, and preventing worldwide epidemics, all within flat or decreasing budgets. As a consequence of the priorities and politics of the time, basic research is too blithely given short shrift, its budget often ending up as the remainder of a growing series of subtractions.

Consequently, success rates in grant applications for basic research are plummeting across all disciplines, particularly for early-career researchers. Life scientists can now expect their first National Institutes of Health grants only in their mid-forties. Apart from discouraging the next generation of talented scholars, this lack of opportunities has led to a much more outcome-driven approach to funding, with granting institutions less willing to place risky long bets. The growing culture of public accountability puts pressure on the margins of error, eliminating downside but also upside risk. The "metrics" used to assess the quality and impact of research proposals—even in the absence of a broadly accepted framework for such measurements—systematically undercut pathbreaking scholarship in favor of more predictable goal-directed research. This number fetishism comes at a great cost, particularly for the humanities and social sciences, whose subtle, complex values and insights easily become invisible when viewed in this harsh quantitative light.

Indeed, in today's metric- and goal-fixated culture, how can we meaningfully convey the "usefulness of useless knowledge"? How many points must one count along research's lengthy, circuitous, and surprising path, often with many dead ends and hairpin turns that lead to further unexpected vistas? How does one articulate a potential outcome of an idea without at the same time boxing it in? Flexner himself admitted that basic research would inevitably waste some money but that the successes would far outweigh the failures. There is no direct or predictable link between the quality of basic research and its effect. The time scales can be long, much longer than the four-year periods in which governments and corporations nowadays tend to think, let alone the twenty-four-hour news cycle. It can easily take many years, even decades, or sometimes, as in the case of Einstein's theory of relativity, a century, for the full societal value of an idea to come to light.

Even if there is a demonstrable correlation with direct applications, it might be difficult to point to an unambiguous causal chain. Applications sometimes appear alongside pure research, and it is not entirely clear who influenced whom, given the many interactions. Attribution is problematic since the stream of knowledge not only meanders, but also branches out, like a river delta flowing into the sea. This challenge of determining origins and outcomes is not confined to scholarship; it is a common human struggle. We are notoriously bad at remembering when, where, and by whom a certain thought pattern was triggered. Since knowledge easily crosses nations and disciplines, this is also true when trying to directly attribute success to specific centers and fields. In the end, science is a truly global, even universal, enterprise.

Finally, we should always remember that basic research does not automatically lead to positive outcomes. Science and technology always bring both lights and shadows. New knowledge can be used for the benefit and the detriment of humankind. This is as true for present-day gene editing methods as for the nuclear technology of Flexner's time. In his essay, Flexner is careful to mention the ambiguous role of scientists in creating new weapons, but he attributes the destructive use of research to "the folly of man, not the intention of scientists." He could not have foreseen the enormity of the atomic bomb that Robert Oppenheimer, the third director of the Institute for Advanced Study, would help to develop in the years directly following the publication of the essay. In the same month that Flexner's article appeared, speaking about loosening "the boundless energy imprisoned in the atom," the first meeting was held of the presidential Advisory Committee on Uranium, a direct result of Einstein's letter to President Roosevelt, that would soon lead to the Manhattan Project and the destruction of Hiroshima and Nagasaki.

* * *

As the 1939 World's Fair demonstrated, it is a difficult, almost impossible task to correctly imagine the world of tomorrow. The Duke of Wellington, conqueror of Napoleon at Waterloo, famously said, "All the business of war, and indeed all the business of life, is to endeavor to find out what you don't know by what you do; that's what I called 'guessing what was at the other side of the hill.'" Indeed, in his retirement, the Iron Duke would take guests to an unknown part of the countryside, where they would try to guess the landscape behind the next hill. Wellington apparently was very good at this game.

Imagination is the ability to see beyond the hill to the unknown side. Curiosity is the innate human urge to climb over the hill and find out what's there. Millions of years of evolutionary payoff have shaped our brains so that we are rewarded for that risky behavior. Recently, neuroscientists have uncovered some of the dopamine activation cycles that stimulate us to venture into unknown territory.

What landscape lies behind the hill? What is left to be discovered? Cosmologists are fortunate in that they think they know exactly what they don't: currently about 95 percent of the universe consists of a mysterious dark matter—five times more abundant than the ordinary matter out of which we, the planets, and the stars are made—and the even more mysterious dark energy, permeating all of empty space. What is the percentage of dark matter in other areas of current knowledge? And, just as interesting, what will the passing of time reveal about what we think we know and what we think we don't know? I would suggest that some of these future answers hinge on how much our society and future societies value and support basic research.

Fundamental science is confronted with an uphill struggle to convince the general public that it's worthy of its support. Such commitment will only materialize if people broadly understand the added purpose and value of looking at the world through the lens of scholarship. Nobody is in a better position to convey that purpose and value than the working scientists and scholars themselves, as they experience the thrill and excitement of research every day in their laboratories, studies, and classrooms.

Therefore, it behooves scientists who want to take practical steps toward improving public support of science to reach out and convey what is exciting about the frontiers that are currently being explored. In this age of unprecedented digital connectivity and communication tools, there are fewer excuses than ever for not informing and engaging the public, and sharing the latest discoveries and personal stories.

We should be encouraged by the public fascination with the big questions that science raises, however far removed from everyday concerns. How did the universe begin and how does it end? What is the origin of life on Earth and possibly elsewhere in the cosmos? What in our brain makes us conscious and human? What will the world of tomorrow bring? Curiosity and imagination are profound qualities that we share with all of humankind.

Einstein opened his address at the 1939 World's Fair with a passionate call for such a public engagement of science: "If science, like art, is to perform its mission totally and fully, its achievements must enter not only superficially but with their inner meaning, into the consciousness of people." Einstein more or less invented the idea of a scientist as a public intellectual, not shy to widely share his scientific insights and commentary on world affairs, spending as much care crafting to perfection his remarks and aphorisms as his mathematical equations.

Flexner was not so comfortable with this highly visible public role, believing that scholars flourish best in isolation. When Einstein settled in Princeton in 1933, President Roosevelt immediately invited America's most famous immigrant to the White House. Flexner intercepted the invitation and declined on Einstein's behalf, writing, "Professor Einstein has come to Princeton for the purpose of carrying on his scientific work in seclusion, and it is absolutely impossible to make any exception which would inevitably bring him into public notice." After this incident, Einstein made sure he personally answered all of his mail.

A broad-ranging dialogue between science and society is not only necessary for laying the foundation for future financial support. It is crucial for attracting young minds to join the research effort. As I have argued before, widely shared knowledge is also a fertile ground for future technology, innovation, and economic growth. Well-informed, science-literate citizens are better able to make responsible choices when confronted with "wicked problems," such as climate change, nuclear power, vaccinations, and genetically modified foods. Similarly, scientists need the dialogue with society to act responsibly in developing potentially harmful technologies. And there is an even higher goal for the public engagement of science: society fundamentally benefits from embracing the scientific culture of accuracy, truth seeking, critical questioning, healthy skepticism, respect for facts and uncertainties, and wonder at the richness of nature and the human spirit.

* * *

The physicist John Wheeler, well-known for his work on nuclear physics and black holes, liked to draw a picture of the universe as a huge capital letter U. On one of its legs, he would draw an eye. We are the universe's eye. Humankind looks at the cosmos, including itself, and, as one of its fortunate observers, applauds the spectacle. Beauty is in the eye of the beholder, and the beauty of the world and the mind is within us. Experiments and equations, theories and telescopes, libraries and laboratories didn't fall out of the sky. They are all hand-made here on planet Earth. We are living in an extraordinary age in which human intelligence, curiosity, and intrepidness are connecting the scientific dots at an exponential rate, enabling a deeper view into the riddle of our existence.

Flexner writes eloquently about how fearless thinking helps to answer fundamental questions about nature and identity: Who am I? Where am I? What does it mean to be a human being? Freedom of thought is essential to human welfare, not only as a tool for advancing knowledge, but also as crucial elements of democracy and tolerance. Like the arts, unfettered scholarship uplifts the spirits, heightens our perspective above the everyday, and shows us a new way to look at the familiar. It literally changes our world. In Flexner's words, "The real enemy of the human race is not the fearless and irresponsible thinker, be he right or wrong. The real enemy is the man who tries to mold the human spirit so that it will not dare to spread its wings."

The World's Fair looked further ahead than dishwashers and television sets. The very long-term future was addressed by a time capsule, buried in the fairgrounds, resistant to corrosion, to be opened only in five thousand years. Apart from a Mickey Mouse cup, copies of *Life* magazine, an array of coins, and many other everyday goods, it contained a letter by Einstein about the human progress and failings of his time, to be read by posterity with, he hoped, "a feeling of proud and justified superiority." His first sentence expresses the promise that scholarship can bring to humanity, which so much infused the fair: "Our time is rich in inventive minds, the inventions of which could facilitate our lives considerably."

In many ways, Flexner's essay can be seen as a similar time capsule, written in a time of great upheaval and anxiety, but fundamentally positive in its long-term outlook. Looking back, it is remarkable how relevant and timely his observations about the power of human curiosity are to the world of today. It's not hard to imagine that this will also be true for the world of tomorrow.

The
Usefulness
of Useless
Knowledge

ABRAHAM FLEXNER

I

Is it not a curious fact that in a world steeped in irrational hatreds which threaten civilization itself, men and women—old and young—detach themselves wholly or partly from the angry current of daily life to devote themselves to the cultivation of beauty, to the extension of knowledge, to the cure of disease, to the amelioration of suffering, just as though fanatics were not simultaneously engaged in spreading pain, ugliness, and suffering? The world has always been a sorry and confused sort of place—yet poets and artists and scientists have ignored the factors that would, if attended to, paralyze them. From a practical point of view, intellectual and spiritual life is, on the surface, a useless form of activity, in which men indulge because they procure for themselves greater satisfactions than are otherwise obtainable. In this paper I shall concern myself with the question of the extent to which the pursuit of these useless satisfactions proves unexpectedly the source from which undreamed-of utility is derived.

We hear it said with tiresome iteration that ours is a materialistic age, the main concern of which should be the wider distribution of material goods and worldly opportunities. The justified outcry of those who through no fault of their own are deprived of opportunity and a fair share of worldly goods therefore diverts an increasing number of students from the studies which their fathers pursued to the equally important and no less urgent study of social, economic, and governmental problems. I have no quarrel with this tendency. The world in which we live is the only world about which our senses can testify. Unless it is made a better world, a fairer world, millions will continue to go to their graves silent, saddened, and embittered. I have myself spent many years pleading that our schools should become more acutely aware of the world in which their pupils and students are destined to pass their lives. Now I sometimes wonder whether that current has not become too strong and whether there would be sufficient opportunity for a full life if the world were emptied of some of the useless things that give it spiritual significance; in other words, whether our conception of what is useful may not have become too narrow to be adequate to the roaming and capricious possibilities of the human spirit.

We may look at this question from two points of view: the scientific and the humanistic or spiritual. Let us take the scientific first. I recall a conversation which I had some years ago with Mr. George Eastman on the subject of use. Mr. Eastman, a wise and gentle farseeing man, gifted with taste in music and art, had been saying to me that he meant to devote his vast fortune to the promotion of education in useful subjects. I ventured to ask him whom he regarded as the most useful worker in science in the world. He replied instantaneously: "Marconi." I surprised him by saying, "Whatever pleasure we derive from the radio or however wireless and the radio may have added to human life, Marconi's share was practically negligible."

I shall not forget his astonishment on this occasion. He asked me to explain. I replied to him somewhat as follows:

"Mr. Eastman, Marconi was inevitable. The real credit for everything that has been done in the field of wireless belongs, as far as such fundamental credit can be definitely assigned to anyone, to Professor Clerk Maxwell, who in 1865 carried out certain abstruse and remote calculations in the field of magnetism and electricity. Maxwell reproduced his abstract equations in a treatise published in 1873. At the next meeting of the British Association, Professor H.J.S. Smith of Oxford declared that 'no mathematician can turn over the pages of these volumes without realizing that they contain a theory which has already added largely to the methods and resources of pure mathematics.' Other discoveries supplemented Maxwell's theoretical work during the next fifteen years. Finally in 1887 and 1888 the scientific problem still remaining—the detection and demonstration of the electromagnetic waves which are the carriers of wireless signals—was solved by Heinrich Hertz, a worker in Helmholtz's laboratory in Berlin. Neither Maxwell nor Hertz had any concern about the utility of their work; no such thought ever entered their minds. They had no practical objective. The inventor in the legal sense was of course Marconi, but what did Marconi invent? Merely the last technical detail, mainly the now obsolete receiving device called coherer, almost universally discarded."

Hertz and Maxwell could invent nothing, but it was their useless theoretical work which was seized upon by a clever technician and which has created new means for communication, utility, and amusement by which men whose merits are relatively slight have obtained fame and earned millions. Who were the useful men? Not Marconi, but Clerk Maxwell and Heinrich Hertz. Hertz and Maxwell were geniuses without thought of use. Marconi was a clever inventor with no thought but use.

The mention of Hertz's name recalled to Mr. Eastman the Hertzian waves, and I suggested that he might ask the physicists of the University of Rochester precisely what Hertz and Maxwell had done; but one thing I said he could be sure of, namely, that they had done their work without thought of use and that throughout the whole history of science most of the really great discoveries which had ultimately proved to be beneficial to mankind had been made by men and women who were driven not by the desire to be useful but merely the desire to satisfy their curiosity.

"Curiosity?" asked Mr. Eastman.

"Yes," I replied, "curiosity, which may or may not eventuate in something useful, is probably the outstanding characteristic of modern thinking. It is not new. It goes back to Galileo, Bacon, and to Sir Isaac Newton, and it must be absolutely unhampered. Institutions of learning should be devoted to the cultivation of curiosity, and the less they are deflected by considerations of immediacy of application, the more likely they are to contribute not only to human welfare but to the equally important satisfaction of intellectual interest which may indeed be said to have become the ruling passion of intellectual life in modern times."

II

What is true of Heinrich Hertz working quietly and unnoticed in a corner of Helmholtz's laboratory in the later years of the nineteenth century may be said of scientists and mathematicians the world over for several centuries past. We live in a world that would be helpless without electricity. Called upon to mention a discovery of the most immediate and far-reaching practical use, we might well agree upon electricity. But who made the fundamental discoveries out of which the entire electrical development of more than one hundred years has come?

The answer is interesting. Michael Faraday's father was a blacksmith; Michael himself was apprenticed to a bookbinder. In 1812, when he was already twenty-one years of age, a friend took him to the Royal Institution, where he heard Sir Humphry Davy deliver four lectures on chemical subjects. He kept notes and sent a copy of them to Davy. The very next year, 1813, he became an assistant in Davy's laboratory, working on chemical problems. Two years later he accompanied Davy on a trip to the Continent. In 1825, when he was thirty-four years of age, he became Director of the Laboratory of the Royal Institution, where he spent fifty-four years of his life.

Faraday's interest soon shifted from chemistry to electricity and magnetism, to which he devoted the rest of his active life. Important but puzzling work in this field had been previously accomplished by Oersted, Ampère, and Wollaston. Faraday cleared away the difficulties which they had left unsolved and by 1841 had succeeded in the task of induction of the electric current. Four years later a second and equally brilliant epoch in his career opened when he discovered the effect of magnetism on polarized light. His earlier discoveries have led to the infinite number of practical applications by means of which electricity has lightened the burdens and increased the opportunities of modern life. His later discoveries have thus far been less prolific of practical results. What difference did this make to Faraday? Not the least. At no period of his unmatched career was he interested in utility. He was absorbed in disentangling the riddles of the universe, at first chemical riddles, in later periods, physical riddles. As far as he cared, the question of utility was never raised. Any suspicion of utility would have restricted his restless curiosity. In the end, utility resulted, but it was never a criterion to which his ceaseless experimentation could be subjected.

In the atmosphere which envelops the world today, it is perhaps timely to emphasize the fact that the part played by science in making war more destructive and more horrible was an unconscious and unintended by-product of scientific activity. Lord Rayleigh, president of the British Association for the Advancement of Science, in a recent address points out in detail how the folly of man, not the intention of the scientists, is responsible for the destructive use of the agents employed in modern warfare. The innocent study of the chemistry of carbon com-

pounds, which has led to infinite beneficial results, showed that the action of nitric acid on
substances like benzene, glycerine, cellulose,
etc., resulted not only in the beneficent aniline
dye industry but in the creation of nitroglycerine, which has uses good and bad. Somewhat
later Alfred Nobel, turning to the same subject,
showed that by mixing nitroglycerine with other
substances, solid explosives which could be
safely handled could be produced—among others, dynamite. It is to dynamite that we owe our
progress in mining, in the making of such rail-
road tunnels as those which now pierce the Alps
and other mountain ranges; but of course dynamite has been abused by politicians and sol-
diers. Scientists are, however, no more to blame
than they are to blame for an earthquake or a
flood. The same thing can be said of poison gas.
Pliny was killed by breathing sulfur dioxide in
the eruption of Vesuvius almost two thousand
years ago. Chlorine was not isolated by scien-

tists for warlike purposes, and the same is true of mustard gas. These substances could be limited to beneficent use, but when the airplane was perfected, men whose hearts were poisoned and whose brains were addled perceived that the airplane, an innocent invention, the result of long disinterested and scientific effort, could be made an instrument of destruction, of which no one had ever dreamed and at which no one had ever deliberately aimed.

In the domain of higher mathematics, almost innumerable instances can be cited. For example, the most abstruse mathematical work of the eighteenth and nineteenth centuries was the "Non-Euclidian Geometry." Its inventor, Gauss, though recognized by his contemporaries as a distinguished mathematician, did not dare to publish his work on "Non-Euclidian Geometry" for a quarter of a century. As a matter of fact, the theory of relativity itself with all its infinite practical bearings would have been utterly impossible without the work which Gauss did at Göttingen.

Again, what is known now as "group theory" was an abstract and inapplicable mathematical theory. It was developed by men who were curious and whose curiosity and puttering led them into strange paths; but "group theory" is today the basis of the quantum theory of spectroscopy, which is in daily use by people who have no idea as to how it came about.

The whole calculus of probability was discovered by mathematicians whose real interest was the rationalization of gambling. It has failed of the practical purpose at which they aimed, but it has furnished a scientific basis for all types of insurance, and vast stretches of nineteenth century physics are based upon it.

From a recent number of *Science* I quote the following:

The stature of Professor Albert Einstein's genius reached new heights when it was disclosed that the learned mathematical physicist developed mathematics fifteen years ago which are now helping to solve the mysteries of the amazing fluidity of helium near the absolute zero of the temperature scale. Before the symposium on intermolecular action of the American Chemical Society Professor F. London, of the University of Paris, now visiting professor at Duke University, credited Professor Einstein with the concept of an "ideal" gas which appeared in papers published in 1924 and 1925.

The Einstein 1925 reports were not about relativity theory, but discussed problems seemingly without any practical significance at the time. They described the degeneracy of an "ideal" gas near the lower limits of the scale of temperature. Because all gases were known to be condensed to liquids at the temperatures in question, scientists rather overlooked the Einstein work of fifteen years ago.

However, the recently discovered behavior of liquid helium has brought the side-tracked Einstein concept to new usefulness. Most liquids increase in viscosity, become stickier and flow less easily, when they become colder. The phrase "colder than molasses in January" is the layman's concept of viscosity and a correct one.

Liquid helium, however, is a baffling exception. At the temperature known as the "delta" point, only 2.19 degrees above absolute zero, liquid helium flows better than it does at higher temperatures and, as a matter of fact, the liquid helium is about as nebulous as a gas. Added puzzles in its strange behavior include its enormous ability to conduct heat. At the delta point it is about 500 times as effective in this respect as copper at room temperature. Liquid helium, with these and other anomalies, has posed a major mystery for physicists and chemists.

Professor London stated that the interpretation of the behavior of liquid helium can best be explained by considering it as a Bose-Einstein "ideal" gas, by using the mathematics worked out in 1924–25, and by taking over also some of the concepts of the electrical conduction of metals. By simple analogy, the amazing fluidity of liquid helium can be partially explained by picturing the fluidity as something akin to the wandering of electrons in metals to explain electrical conduction.

Let us look in another direction. In the domain of medicine and public health, the science of bacteriology has played for half a century the leading role. What is its story? Following the Franco-Prussian War of 1870, the German government founded the great University of Strasbourg. Its first professor of anatomy was Wilhelm von Waldeyer, subsequently professor of anatomy in Berlin. In his *Reminiscences* he relates that among the students who went with him to Strasbourg during his first semester there was a small, inconspicuous, self-contained youngster of seventeen by name Paul Ehrlich. The usual course in anatomy then consisted of dissection and microscopic examination of tissues. Ehrlich paid little or no attention to dissection, but, as Waldeyer remarks in his *Reminiscences*:

> I noticed quite early that Ehrlich would work long hours at his desk, completely absorbed in microscopic observation. Moreover, his desk gradually became covered with colored spots of every description. As I saw him sitting at work one day, I went

up to him and asked what he was doing with all his rainbow array of colors on his table. Thereupon this young student in his first semester supposedly pursuing the regular course in anatomy looked up at me and blandly remarked, *"Ich probiere."* This might be freely translated, "I am trying" or "I am just fooling." I replied to him, "Very well. Go on with your fooling." Soon I saw that without any teaching or direction whatsoever on my part I possessed in Ehrlich a student of unusual quality.

Waldeyer wisely left him alone. Ehrlich made his way precariously through the medical curriculum and ultimately procured his degree mainly because it was obvious to his teachers that he had no intention of ever putting his medical degree to practical use. He went subsequently to Breslau, where he worked under Professor Cohnheim, the teacher of our own Dr.

Welch, founder and maker of the Johns Hopkins Medical School. I do not suppose that the idea of use ever crossed Ehrlich's mind. He was interested. He was curious; he kept on fooling. Of course his fooling was guided by a deep instinct, but it was a purely scientific, not a utilitarian, motivation. What resulted? Koch and his associates established a new science, the science of bacteriology. Ehrlich's experiments were now applied by a fellow student, Weigert, to staining bacteria and thereby assisting in their differentiation. Ehrlich himself developed the staining of the blood film with the dyes on which our modern knowledge of the morphology of the blood corpuscles, red and white, is based. Not a day passes but that in thousands of hospitals the world over Ehrlich's technic is employed in the examination of the blood. Thus the apparently aimless fooling in Waldeyer's dissecting room in Strasbourg has become a main factor in the daily practice of medicine.

I shall give one example from industry, one selected at random, for there are scores besides. Professor Berl, of the Carnegie Institute of Technology (Pittsburgh) writes as follows:

The founder of the modern rayon industry was the French Count Chardonnet. It is known that he used a solution of nitro cotton in ether-alcohol, and that he pressed this viscous solution through capillaries into water which served to coagulate the cellulose nitrate filament. After the coagulation, this filament entered the air and was wound up on bobbins. One day Chardonnet inspected his French factory at Besançon. By an accident the water which should coagulate the cellulose nitrate filament was stopped. The workmen found that the spinning operation went much better without water than with water. This was the birthday of the very important process of dry spinning, which is actually carried out on the greatest scale.

III

I am not for a moment suggesting that everything that goes on in laboratories will ultimately turn to some unexpected practical use or that an ultimate practical use is its actual justification. Much more am I pleading for the abolition of the word "use," and for the freeing of the human spirit. To be sure, we shall thus free some harmless cranks. To be sure, we shall thus waste some precious dollars. But what is infinitely more important is that we shall be striking the shackles off the human mind and setting it free for the adventures which in our own day have, on the one hand, taken Hale and Rutherford and Einstein and their peers millions upon millions of miles into the uttermost realms of space and, on the other, loosed the boundless energy imprisoned in the atom. What Rutherford and others like Bohr and Millikan have done out of sheer curiosity in the effort to understand the construction of the atom has released forces which may transform human life; but this ultimate and unforeseen and unpredictable practical result is not offered as a justification for Rutherford or Einstein or Millikan or Bohr or any of their peers. Let them alone. No educational administrator can possibly direct the channels in which these or other men shall work. The waste, I admit again, looks prodigious. It is not really so. All the waste that could be summed up in developing the science of bacteriology is as nothing compared to the advantages which have accrued from the discoveries of Pasteur, Koch, Ehrlich, Theobald Smith, and scores of others—advantages that could never have accrued if the idea of possible use had permeated their minds. These great artists—for such are scientists and bacteriologists—disseminated the spirit which prevailed in laboratories in which they were simply following the line of their own natural curiosity.

I am not criticising institutions like schools of engineering or law in which the usefulness motive necessarily predominates. Not infrequently the tables are turned, and practical difficulties encountered in industry or in laboratories stimulate theoretical inquiries which may or may not solve the problems by which they were suggested, but may also open up new vistas, useless at the moment, but pregnant with future achievements, practical and theoretical.

With the rapid accumulation of "useless" or theoretic knowledge, a situation has been created in which it has become increasingly possible to attack practical problems in a scientific spirit. Not only inventors, but "pure" scientists have indulged in this sport. I have mentioned Marconi, an inventor, who, while a benefactor to the human race, as a matter of fact merely "picked other men's brains." Edison belongs to the same category. Pasteur was different. He was a great scientist; but he was not averse to attacking practical problems—such as the condition of French grapevines or the problems of beer-brewing—and not only solving the immediate difficulty, but also wresting from the practical problem some far-reaching theoretic conclusion, "useless" at the moment, but likely in some unforeseen manner to be "useful" later. Ehrlich, fundamentally speculative in his curiosity, turned fiercely upon the problem of syphilis and doggedly pursued it until a solution of immediate practical use—the discovery of salvarsan—was found. The discoveries of insulin by Banting for use in diabetes and of liver extract by Minot and Whipple for use in pernicious anemia belong in the same category: both were made by thoroughly scientific men, who realized that much "useless" knowledge had been piled up by men unconcerned with its practical bearings, but that the time was now ripe to raise practical questions in a scientific manner.

Thus it becomes obvious that one must be wary in attributing scientific discovery wholly to any one person. Almost every discovery has a long and precarious history. Someone finds a bit here, another a bit there. A third step succeeds later and thus onward till a genius pieces the bits together and makes the decisive contribution. Science, like the Mississippi, begins in a tiny rivulet in the distant forest. Gradually other streams swell its volume. And the roaring river that bursts the dikes is formed from countless sources.

I cannot deal with this aspect exhaustively, but I may in passing say this: over a period of one or two hundred years the contributions of professional schools to their respective activities will probably be found to lie, not so much in the training of men who may tomorrow become practical engineers or practical lawyers or practical doctors, but rather in the fact that even in the pursuit of strictly practical aims an enormous amount of apparently useless activity goes on. Out of this useless activity there come discoveries which may well prove of infinitely more importance to the human mind and to the human spirit than the accomplishment of the useful ends for which the schools were founded.

The considerations upon which I have touched emphasize—if emphasis were needed—the overwhelming importance of spiritual and intellectual freedom. I have spoken of experimental science; I have spoken of mathematics; but what I say is equally true of music and art and of every other expression of the untrammeled human spirit. The mere fact that they bring satisfaction to an individual soul bent upon its own purification and elevation is all the justification that they need. And in justifying these without any reference whatsoever, implied or actual, to usefulness, we justify colleges, universities, and institutes of research. An institution which sets free successive generations of human souls is amply justified whether or not this graduate or that makes a so-called useful contribution to human knowledge. A poem, a symphony, a painting, a mathematical truth, a new scientific fact, all bear in themselves all the justification that universities, colleges, and institutes of research need or require.

The subject which I am discussing has at this moment a peculiar poignancy. In certain large areas—Germany and Italy especially—the effort is now being made to clamp down the freedom of the human spirit. Universities have been so reorganized that they have become tools of those who believe in a special political, economic, or racial creed. Now and then a thoughtless individual in one of the few democracies left in this world will even question the fundamental importance of absolutely untrammeled academic freedom. The real enemy of the human race is not the fearless and irresponsible thinker, be he right or wrong. The real enemy is the man who tries to mold the human spirit so that it will not dare to spread its wings, as its wings were once spread in Italy and Germany, as well as in Great Britain and the United States.

This is not a new idea. It was the idea which animated von Humboldt when, in the hour of Germany's conquest by Napoleon, he conceived and founded the University of Berlin. It is the idea which animated President Gilman in the founding of the Johns Hopkins University, after which every university in this country has sought in greater or less degree to remake itself. It is the idea to which every individual who values his immortal soul will be true whatever the personal consequences to himself. Justification of spiritual freedom goes, however, much farther than originality whether in the realm of science or humanism, for it implies tolerance throughout the range of human dissimilarities. In the face of the history of the human race, what can be more silly or ridiculous than likes or dislikes founded upon race or religion? Does humanity want symphonies and paintings and profound scientific truth, or does it want Christian symphonies, Christian paintings, Christian science, or Jewish symphonies, Jewish paintings, Jewish science, or Mohammedan or Egyptian or Japanese or Chinese or American or German or Russian or Communist or Conservative contributions to and expressions of the infinite richness of the human soul?

IV

Among the most striking and immediate consequences of foreign intolerance I may, I think, fairly cite the rapid development of the Institute for Advanced Study, established by Mr. Louis Bamberger and his sister, Mrs. Felix Fuld, at Princeton, New Jersey. The founding of the Institute was suggested in 1930. It was located at Princeton partly because of the founders' attachment to the State of New Jersey, but, in so far as my judgment was concerned, because Princeton had a small graduate school of high quality with which the most intimate cooperation was feasible. To Princeton University the Institute owes a debt that can never be fully appreciated. The work of the Institute with a considerable portion of its personnel began in 1933. On its faculty are eminent American scholars—Veblen, Alexander, and Morse, among the mathematicians; Meritt, Lowe, and Miss Goldman among the humanists; Stewart, Riefler, Warren, Earle, and Mitrany among the publicists and economists. And to these should be added scholars and scientists of equal caliber already assembled in Princeton University, Princeton's library, and its laboratories. But the Institute for Advanced Study is indebted to

Hitler for Einstein, Weyl, and von Neumann in mathematics; for Herzfeld and Panofsky in the field of humanistic studies, and for a host of younger men who during the past six years have come under the influence of this distinguished group and are already adding to the strength of American scholarship in every section of the land.

The Institute is, from the standpoint of organization, the simplest and least formal thing imaginable. It consists of three schools—a School of Mathematics, a School of Humanistic Studies, a School of Economics and Politics. Each school is made up of a permanent group of professors and an annually changing group of members. Each school manages its own affairs as it pleases; within each group each individual disposes of his time and energy as he pleases. The members who already have come from twenty-two foreign countries and thirty-nine institutions of higher learning in the

United States are admitted, if deemed worthy, by the several groups. They enjoy precisely the same freedom as the professors. They may work with this or that professor, as they severally arrange; they may work alone, consulting from time to time anyone likely to be helpful. No routine is followed; no lines are drawn between professors, members, or visitors. Princeton students and professors and Institute members and professors mingle so freely as to be indistinguishable. Learning as such is cultivated. The results to the individual and to society are left to take care of themselves. No faculty meetings are held; no committees exist. Thus men with ideas enjoy conditions favorable to reflection and to conference. A mathematician may cultivate mathematics without distraction; so may a humanist in his field, an economist or a student of politics in his. Administration has been minimized in extent and importance. Men without ideas, without power of concentration on ideas, would not be at home in the Institute.

I can perhaps make this point clearer by citing briefly a few illustrations. A stipend was awarded to enable a Harvard professor to come to Princeton: he wrote asking,

"What are my duties?"

I replied, "You have no duties—only opportunities."

An able young mathematician, having spent a year at Princeton, came to bid me good-by. As he was about to leave, he remarked:

"Perhaps you would like to know what this year has meant to me."

"Yes," I answered.

"Mathematics," he rejoined, "is developing rapidly; the current literature is extensive. It is now over ten years since I took my Ph.D. degree. For a while I could keep up with my subject; but latterly that has become increasingly difficult and uncertain. Now, after a year here, the blinds are raised; the room is light; the windows are open. I have in my head two papers that I shall shortly write."

"How long will this last?" I asked.

"Five years, perhaps ten."

"Then what?"

"I shall come back."

A third example is of recent occurrence. A professor in a large Western university arrived in Princeton at the end of last December. He had in mind to resume some work with Professor Morey (at Princeton University). But Morey suggested that he might find it worthwhile to see Panofsky and Swarzenski (at the Institute). Now he is busy with all three.

"I shall stay," he added, "until next October."

"You will find it hot in midsummer," I said.

"I shall be too busy and too happy to notice it."

Thus freedom brings not stagnation, but

rather the danger of overwork. The wife of an English member recently asked:

"Does everyone work until two o'clock in the morning?"

The Institute has had thus far no building. At this moment the mathematicians are guests of the Princeton mathematicians in Fine Hall; some of the humanists are guests of the Princeton humanists in McCormick Hall; others work in rooms scattered through the town. The economists now occupy a suite at The Princeton Inn. My own quarters are located in an office building on Nassau Street, where I work among shopkeepers, dentists, lawyers, chiropractors, and groups of Princeton scholars conducting a local government survey and a study of population. Bricks and mortar are thus quite inessential, as President Gilman proved in Baltimore sixty-odd years ago. Nevertheless, we miss informal contact with one another and are about to remedy this defect by the erection of a building provided by the founders, to be called Fuld Hall. But formality shall go no farther. The Institute must remain small; and it will hold fast to the conviction that The Institute Group desires leisure, security, freedom from organization and routine, and, finally, informal contacts with the scholars of Princeton University and others who from time to time can be lured to Princeton from distant places. Among these, Niels Bohr has come from Copenhagen, von Laue from Berlin, Levi-Civita from Rome, André Weil from Strasbourg, Dirac and G. H. Hardy from Cambridge, Pauli from Zurich, Lemaître from Louvain, Wade-Gery from Oxford, and Americans from Harvard, Yale, Columbia, Cornell, Johns Hopkins, Chicago, California, and other centers of light and learning.

We make ourselves no promises, but we cherish the hope that the unobstructed pursuit of

useless knowledge will prove to have consequences in the future as in the past. Not for a
moment, however, do we defend the Institute
on that ground. It exists as a paradise for scholars who, like poets and musicians, have won the
right to do as they please and who accomplish
most when enabled to do so.

About the Authors

ROBBERT DIJKGRAAF, director and Leon Levy Professor of the Institute for Advanced Study in Princeton, is a mathematical physicist who has made significant contributions to string theory and the advancement of science education. Dijkgraaf is president of the InterAcademy Partnership, past president of the Royal Netherlands Academy of Arts and Sciences, and a distinguished public policy adviser and advocate for science and the arts.

ABRAHAM FLEXNER conceived of and developed the Institute for Advanced Study, serving as its founding director from 1930 to 1939. A prominent figure in U.S. education reform, including medical training and practice, Flexner had a profound impact on the advancement of knowledge and academic integrity and ideals. The Institute, dedicated to curiosity-driven basic research, remains at the heart of his legacy.

Further Reading

Alberts, Bruce, Marc W. Kirschner, Shirley Tilghman, and Harold Varmus. "Rescuing U.S. Biomedical Research from Its Systemic Flaws." *Proceedings of the National Academy of Sciences*, Vol. 111, No. 16, April 22, 2014.

Bohr, Niels, and John Archibald Wheeler. "The Mechanism of Nuclear Fission." *Physical Review*, Vol. 56, Issue 5, September 1, 1939.

Bush, Vannevar. *Science, the Endless Frontier*, July 1945 (reprinted July 1960, National Science Foundation, Washington, D.C.).

Einstein's Letter (1939). U.S. Department of Energy, Office of Science and Technical Information, Office of History and Heritage Resources, www.osti.gov/opennet/manhattan-project -history/Events/1939-1942/einstein_letter.htm.

Flexner, Abraham. *The American College: A Criticism.* The Century Co., New York, 1908.

Flexner, Abraham. *I Remember, the Autobiography of Abraham Flexner.* Simon and Schuster, New York, 1940.

Flexner, Abraham. *Medical Education in the United States and Canada.* Carnegie Foundation for the Advancement of Teaching, New York, 1910.

Flexner, Abraham. *Universities: American, English, German.* Oxford University Press, New York, 1930.

Joint Economic Committee. "The Role of Research & Development in Strengthening America's Innovation Economy." U.S. Congress, December 2014.

Massachusetts Institute of Technology. *The Future Postponed: Why Declining Investment in Basic Research Threatens a U.S. Innovation Deficit.* Cambridge, April 2015.

New York Times. "President Opens Fair as a Symbol of Peace; Vast Spectacle of Color and World Progress Thrills Enthusiastic Crowds on the First Day," May 1, 1939.

Report of the Director, Minutes of the Regular Meeting of the Institute for Advanced Study. Shelby White and Leon Levy Archives Center, May 22, 1939.

van 't Hoff, Jacobus Henricus. *De Verbeeldingskracht in de Wetenschap*. P. M. Bazendijk, Rotterdam, Netherlands, 1878. Translated by G. F. Springer as *Imagination in Science* (Springer, 1967).

Wilson, Robert. "R. R. Wilson's Congressional Testimony." April 1969. Fermilab History and Archives Project, http://history.fnal.gov/testi mony.html.

A Note on the Type

This book has been composed in Miller, a Scotch Roman typeface designed by Matthew Carter and first released by Font Bureau in 1997. It bears a striking resemblance to Monticello, the typeface developed for The Papers of Thomas Jefferson in the 1940s by C. H. Griffith and P. J. Conkwright and reinterpreted in digital form by Carter in 2003.